Arts & Crafts

PAPER MAKING

Susie O'Reilly

With photographs by Zul Mukhida

Wayland

Titles in this series

BATIK AND TIE-DYE
BLOCK PRINTING
KNITTING AND CROCHET
MODELLING
NEEDLECRAFT
PAPER MAKING
STENCILS AND SCREENS
WEAVING

Frontispiece *The Japanese use paper to make all kinds of objects. Paper fans are popular.*

This edition published in 1994
by Wayland (Publishers) Ltd

First published in 1993 by
Wayland (Publishers) Ltd
61 Western Road, Hove
East Sussex BN3 1JD, England

Editor: Anna Girling
Designer: Jean Wheeler

British Library Cataloguing in Publication Data
O'Reilly, Susie
Paper Making. – (Arts & Crafts Series)
I. Title II. Series
676

HARDBACK ISBN 0-7502-0969-0

PAPERBACK ISBN 0-7502-1414-7

Typeset by Dorchester Typesetting
Group Ltd, Dorchester, Dorset, England
Printed and bound by Lego, Italy

Contents

Words printed in **bold** appear in the glossary.

Getting Started

We use paper every day, probably without even thinking about it. We write on it, read books made of it, wrap food in it and, more often than not, screw it up and throw it away. But what is paper?

Paper is made from the long, thin threadlike **fibres** that plants and trees are made up of. The fibres are beaten to separate them and then mixed with water to form a **pulp**. The pulp is put on a fine-mesh screen called a **mould**, which strains out the water, leaving a sheet of matted fibres on the mesh. Once it is dry this thin layer of fibres is paper.

This basic method is used in the making of all paper. It is used in industry to make enormous lengths of paper on vast machines. And it is used by individual paper makers producing hand-made sheets in small paper mills or artists' studios.

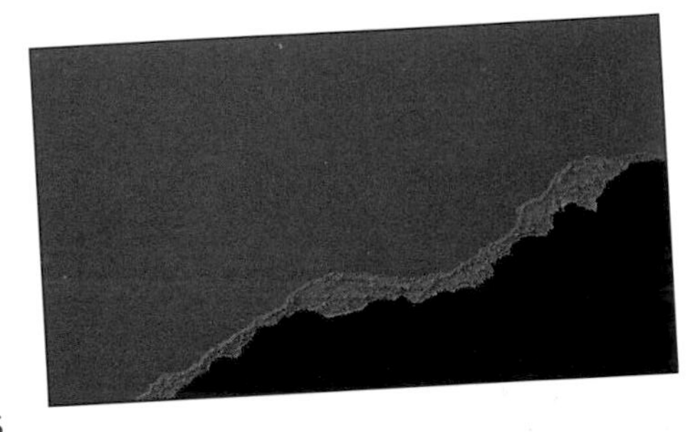

◀ *It is easy to see the fibres in a torn piece of paper.*

The best-quality paper is made from the long, strong fibres of the inner bark of **hardwood** shrubs (such as mulberry bushes) or from **recycled** cotton and **linen** rags. Since the late nineteenth century, however, most of our everyday paper has been made from **softwood** trees such as pines and spruces. These trees are specially grown in forests in places such as Scandinavia, North America and Indonesia.

◀ *Most of our paper is made in paper mills using vast machinery. It is made in rolls, not sheets.*

Making paper by hand using a small mould to form each sheet. ▶

Paper is more than just a plain, white surface to write on. In recent years many artists have found that, rather than simply painting and printing on paper, they can turn the sheet of paper itself into a work of art. Making paper by hand allows plenty of scope for experimenting. Paper can be shaped or coloured, or have other objects added to it. There are all kinds of interesting possibilities.

The Diver, *by the artist David Hockney. Hockney worked at a small paper making studio to produce the sheets for this picture. The design is in the paper itself.* ▶

YOU WILL NEED

To make paper pulp
A variety of waste papers to recycle, e.g. computer paper; paper bags; wrapping paper; tissue paper. (Plastic-coated, shiny or highly illustrated papers are not suitable.)

To make pulp from plants
The leaves and stems of different plants, e.g. rushes; grasses; nettles; the outer leaves of cauliflowers; leaves and stalks of rhubarb; celery stalks; onion peelings; iris leaves.

To make the mould
Nylon mesh (such as curtain netting) with between 10 and 30 holes per cm^2
Stainless steel staples and strong stapler
Rustproof nails
180-cm length of 1 cm × 1 cm wood
A small saw
A hammer
PVA glue

To prepare the pulps
A bucket
A wooden stick
An electric liquidizer
A washing-up bowl
A large saucepan with lid (not aluminium)
An electric or gas ring
Washing soda

To make a series of sheets of paper
A shallow rectangular tray
All-purpose kitchen cloths
Two pieces of hardboard, 25 cm × 20 cm
Two bricks
A large dropper, such as a turkey baster
A flat-bladed knife
Clean, dry newspaper
An electric iron
A rolling pin

To colour the paper
Tea bags
Instant coffee
Food dyes
Powder paints
Coloured papers

General equipment
Apron; rubber gloves; scissors; secateurs; newspaper for protecting surfaces; drawing pins; pinboard.

THE HISTORY OF PAPER MAKING

Around 4,000 years ago, the ancient Egyptians made a kind of paper using the **papyrus** plant. Parts of the stem of the papyrus were pounded until they were flat, laid out to form a layer and covered by muddy water from the River Nile. More, criss-crossing layers were placed on top. Finally the sheet of papyrus was pressed and left in the sun to dry. The ancient Greeks and Romans also used scrolls made from papyrus. The word 'paper' comes from the Greek word *papuros.*

▲ *A piece of papyrus from Egypt, made more than 3,000 years ago. Can you see the criss-cross effect made by the strips of papyrus?*

The discovery of how to make a light, flat, smooth surface to write on was enormously important. However, the knowledge of how to make paper spread around the world only very slowly. This was partly because links between countries were poor, and also because people who knew the technique kept it as a jealously guarded secret.

A picture of early paper making in China. ▼

It was the Chinese who discovered the technique of making paper as we know it today. The accepted date for their invention of paper is AD 105. In fact, at around that time many people in China were experimenting to find a better surface to write on. The Chinese were great **calligraphers**, but found writing on woven cloth or strips of wood and bamboo difficult. Storage was also a problem – bundles of wooden sticks take up a great deal of space.

The Japanese developed the technique of paper making into a highly skilled craft. Japan is still the country most admired for its hand paper making. It is mainly a winter activity, carried out by peasant farmers to bring in extra money. Japanese paper makers use a paper mould with a **flexible** bamboo mesh rather than the rigid mesh that is used in other countries. This allows the fibres to criss-cross more and produces stronger paper.

A Japanese paper maker using a flexible bamboo mesh. ▶

In Europe and North America, until the end of the eighteenth century, all paper was made by hand, mainly from rags. Rag-and-bone men used to visit houses with their carts to buy people's cast-off clothing, which they sold to the paper mills. However, in 1798 the first paper-making machine was invented in France. Machine-made paper was cheap and the demand for it grew. As a result, there were simply not enough rags to supply the **manufacturers**, who started to search for a new material for making into paper. The idea of using wood came from the study of wasps. Female wasps chew up old wood to make sawdust. Then they mix it with their **saliva** to make a pulp and form paper nests to lay their eggs in.

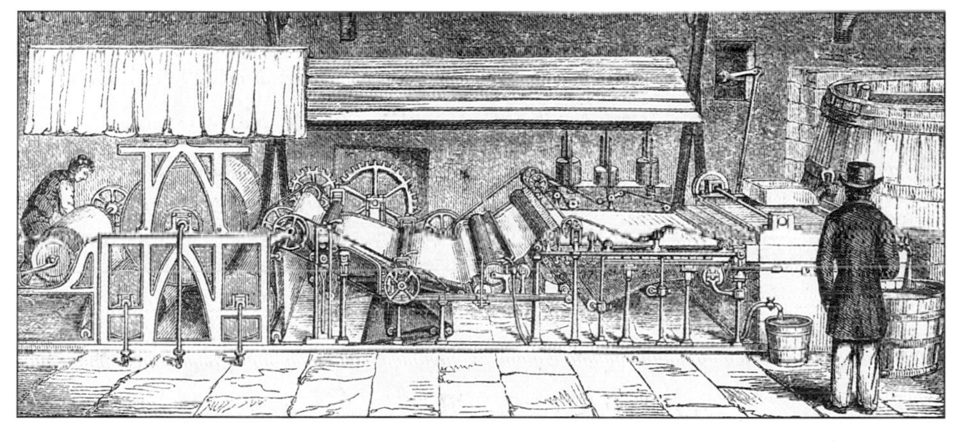

▼ *An early paper-making machine from Europe in the nineteenth century.*

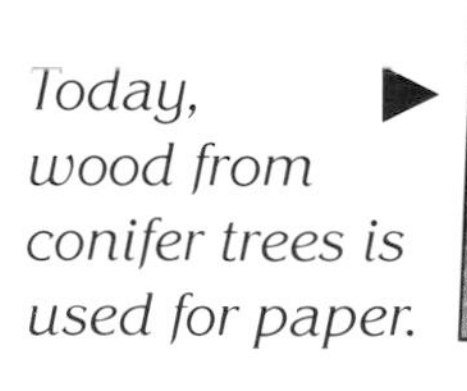

Today, wood from conifer trees is used for paper. ▶

Eventually, it was discovered that fibres from **conifer** trees, such as spruce, pine and fir, could be used to make paper. This led to an enormous growth in the industry. A much greater range of papers, with a wide range of uses, could now be produced cheaply.

Extraordinary Uses of Paper

Paper can be put to an astonishing number of different uses. It was invented because people needed a surface to write on. Writing on paper is quicker and easier than carving words into stone, clay or wax **tablets**. However, paper has also been used – and still is used – in many other ways.

The Japanese use paper almost as much as they use fabrics. Paper is used to make kites, umbrellas, lanterns, fans, doors, windows and luggage. The Japanese even make **woven** paper clothing, called *shifu*. A sheet of paper is cut into strips, which are rolled on a stone to make them into threads. These are then woven on a **loom** into cloth. The cloth can be cut and sewn to make **kimonos**, jackets, purses and other items. Japanese warriors even used paper, mixed with other materials, to make lightweight armour.

▲ *A paper kimono from Japan.*

◀ *In the Far East, paper is used to make umbrellas. This colourful display is in Thailand.*

In the nineteenth century, in both Europe and North America, paper was used instead of wood and metal to make many items. People discovered how to make **laminated** panels by gluing sheets of paper together. These panels were waterproof and heat-resistant. They could be sawn into pieces or moulded into different shapes. They were used to make coffins, wall panels, cupboards, chairs and tables, as well as smaller items such as trays and sewing boxes. These objects were painted and then layers of **varnish** were carefully applied to protect the surface and make it very smooth.

Today, paper is used to make throw-away items. Doctors wear paper gowns and masks when they are carrying out operations. Many babies now wear **disposable** paper nappies instead of cotton nappies. Disposable paper cups, plates and napkins are used by many takeaway restaurants. Most of the things we buy in shops and supermarkets are wrapped in paper packaging.

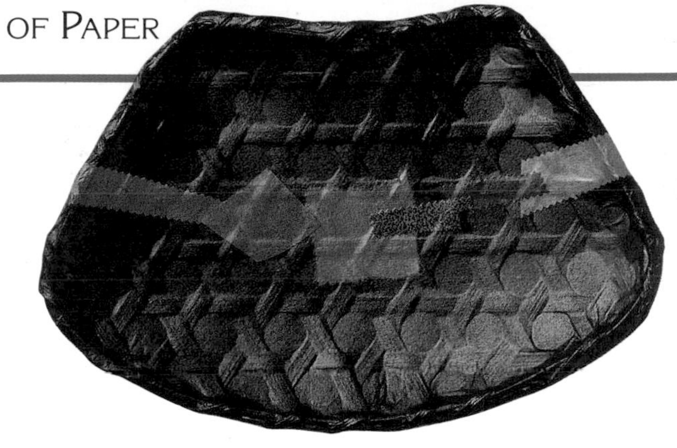

▲ *Have you used paper plates at a picnic? This beautiful plate is from Japan.*

The world is now using up paper faster than new trees can be grown to produce it. However, the plant fibres used to make paper can be reused. **Environmentalists** have been trying to make people see the need to recycle paper in this way. Recycling is essential, not simply to save trees for making paper in the future, but also because trees play an important part in putting oxygen into the Earth's **atmosphere** and controlling **climate**. If we continue to cut down trees, the delicate balance of gases that keeps the Earth's plants, animals and humans alive will be destroyed.

◀ *Paper from offices is collected ready to be treated and made into recycled paper (below).*

MAKING A MOULD

A paper-making mould acts like a sieve. It is a frame on which the sheet of paper is formed from the paper pulp. The mould has a fine mesh stretched across it which catches the fibres in the pulp while allowing the water to drain away. A second frame, the same size and shape as the mould but without the mesh, helps to make the edges of the sheets of paper straight. This second frame is called a deckle.

1 Buy a length of wood that is 1 cm × 1 cm thick. It will need to be cut into four pieces that are 25 cm long and four that are 20 cm long.

2 Arrange two longer and two shorter pieces of wood to make a rectangular frame. Glue the corners together with PVA glue.

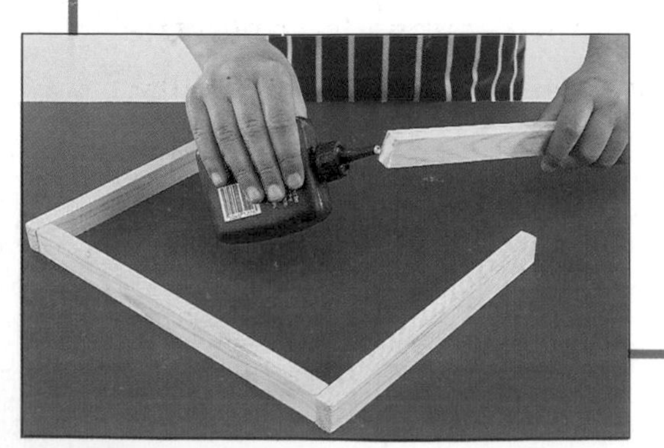

3 When the glue is dry, hammer in two nails at each corner, to hold the sides in place. **Ask an adult to help you.**

4 Find or buy a piece of fine-mesh nylon netting, such as an old net curtain. Cut a piece measuring 30 cm × 25 cm.

5 Wet the netting and staple it to the frame using a large staple gun. Start by stapling the middle of one side, then the middle of the opposite side, pulling as tightly as you can. Then staple the other two sides.

Remember: always be careful with a staple gun and staples. Ask an adult to help you.

6 Now staple all round the frame, making sure the netting is **taut**. As it dries, it will shrink and become tighter. Cut away any extra netting.

7 Brush round the edges and sides of the frame with a waterproof glue, such as PVA. This will neaten the edges of the mould and will help to hold the netting taut.

8 To make the deckle, repeat steps 2–3, using the other four pieces of wood. It must be exactly the same size as the paper mould.

Note: paper making uses a lot of water. It is important to use waterproof glue (such as PVA glue) and nails and staples that will not rust (use ones made of steel, not iron). Rust will mark your paper.

You can also use old picture frames to make the mould and deckle, as long as they are about the same size. The deckle can be a little smaller than the mould, but not larger.

Making Pulp by Recycling Paper

You can recycle almost any kind of paper. Computer print-out paper is excellent – because it needs to be strong, it is made from pulp with long fibres. Paper bags and envelopes are also good for recycling.

Avoid any paper with a shiny surface. Shiny papers are coated with clay. This can cause powdery patches on your sheets. It is fine to use paper that has been printed on, including newspaper. However, newspaper does not make very strong paper because it is made of pulp with short fibres. Recycling the paper makes the fibres even shorter.

1 Prepare the paper by removing all traces of glue and taking out any staples.

2 Tear the paper into small squares, about the size of postage stamps.

3 Put the torn paper in a bucket of water and leave it to soak for at least two hours or, preferably, overnight.

4 Tip away some of the water and use a long, thick piece of wood to beat the paper to a mushy pulp. This can take quite a long time. The pulp needs to be very smooth and creamy.

5 Fill a rectangular plastic tray with water. Pour the pulp into the tray and stir it until it is thoroughly mixed with the water. This is your '**vat**' of pulp, which is now ready to be made into paper.

6 A quicker method of making pulp is to use an electric liquidizer. **Ask an adult to help you**. Put a handful of torn paper in the liquidizer and fill it two-thirds full with water.

Replace the lid and screw it on tightly. Liquidize for 30 seconds. Do this in three bursts of 10 seconds each, resting the machine between bursts. You will need to make a total of three loads in the liquidizer to make a vat of pulp.

Never pour paper pulp down a sink – it will block the drain. When you have finished using a pulp, drain off the water through fine-mesh netting, using your hands to squeeze out all the water. Pour the water down the sink and either keep the dry pulp to use later or throw it away in a rubbish bin.

Making a Sheet of Paper

Paper making uses a lot of water. Take care to avoid getting water everywhere.

1 Take a large newspaper and open it out flat on an old table. Now place a small newspaper in the middle. Place a piece of hardboard on the newspaper. Cover with a damp kitchen cloth.

2 Give your pulp another good stir. Take the mould so that the mesh is uppermost and put the deckle on top. Grip the two together firmly, holding them by the shorter edges.

3 Slip the mould and deckle at an angle into the pulp mixture, from the far side of the vat. Then straighten them up so that they lie flat beneath the surface of the liquid.

4 Keeping the mould and deckle level, pull them straight up out of the liquid. Hold them over the vat so that the water can drain back through the mesh.

5 Gently shake the mould and deckle backwards and forwards and from side to side. This will help the fibres to settle down. Do not overdo it, or your sheet of paper will be uneven.

6 Take away the deckle and transfer the mould to the prepared newspaper pad, turning it over so that the pulp side faces downwards on the kitchen cloth.

7 In one gentle movement, press one edge of the sheet of paper on to the kitchen cloth, lift up the opposite edge of the mould and take it away. The sheet should have been transferred to the kitchen cloth. You have made your first piece of paper!

You may need a few goes before you manage to get an even layer of pulp on the mesh and then turn it out as a single sheet. If things go wrong, you can remove the pulp from the cloth by laying the cloth on top of the pulp in the vat.

Helpful Hints

1 Use a large dropper (a turkey baster is excellent), full of pulp mixture, to fill in any areas where the sheet of paper is thin.

2 Straighten the edges of the sheet of paper by gently pushing the edges with the side of a flat-bladed knife.

Turn to pages 12-13 and 18-19 for information on making paper pulp.

Couching, Sizing and Finishing

Couching

'Couching' is a special word used by paper makers. It describes the process of transferring a sheet of newly made paper on to a drying cloth or board, pressing it and drying it. This is how you make a series of sheets of paper. Try working in pairs – one of you as vat-person and one as coucher.

1 Place a damp kitchen cloth on top of the first sheet of paper. Smooth out any wrinkles and make sure it is flat. Any creases will show up in the sheets of paper you make.

2 Make more sheets of paper and put them on top of one another, placing damp cloths between each. Continue until you have made five sheets.

3 Cover the last sheet with a damp cloth. Take a second piece of hardboard and place it on top. Place two bricks on the board and leave for a quarter of an hour.

4 Put a sheet of plastic over a large, flat surface (a table or an area of floor that people will not walk over). Spread a thick layer of newspaper over the plastic sheet. Remove the top board and carefully peel off each kitchen cloth with its sheet of paper. Do not worry about damaging the sheets – the damp paper is very flexible.

5 Leave the sheets on the newspaper to dry. This will take between twelve and twenty-four hours at least. Change the newspapers whenever they become soaked. Drying the sheets in a warm room will speed up the process.

SIZING

'Sizing' does not mean cutting the paper to the correct size! In paper making, sizing means treating paper to make it strong and **non-absorbent**, so that ink and paint do not soak into the surface and make a blot.

Here are several ways of sizing. Waxed and oiled papers are not suitable for writing on but they are strong and have an interesting surface.

1 Mix a tablespoonful of PVA glue into a jar of water. Stir it thoroughly until all the lumps have gone. Add this to your paper pulp *before* you start making the paper.

2 Paint both sides of a sheet of paper with a thin coat of vegetable oil. This will make it **translucent**.

3 Coat the paper with wax. Put some wax candles in an old tin can and put the can in a saucepan, half full of water. Heat the pan of water on an electric or gas ring until the candles melt. Brush the melted wax over the paper using an old brush.

Ask an adult to help you melt the wax. Do not let the water boil dry.

FINISHING

You can give different **finishes** to your paper. Here are some ideas.

1 If you want your paper to have a rough surface, leave the sheets to dry out completely before gently peeling them away from the cloths.

2 If you want the paper to have a smooth, even finish, roll each sheet with a rolling pin while it is still slightly damp.

3 Dry your paper under weights to make sure it dries absolutely flat. Two bricks placed on a board make good weights.

MAKING PAPER FROM PLANTS

As well as recycling waste paper, it is possible to make paper from plants. Plants with long, narrow leaves, such as rushes, reeds, maize plants and irises, or stringy stalks, such as rhubarb, celery, cow parsley and sunflowers, are particularly suitable.

Try to use plant matter that would otherwise be thrown away. For example, you can use vegetable peelings, such as onion skins and outer cabbage leaves. If you live near a market, you might be able to get some more unusual leaves, such as the tops of pineapples. The plants you use will depend on the time of the year and what is easily available.

Producing paper from plants is slightly more complicated than recycling old paper because the plant matter must be broken down first to release the fibres.

▲ *Paper made from plants: (clockwise from top left) coriander and daffodil petals; pineapple tops; celery and fennel; iris.*

1 Collect your plant matter. You will need half a washing-up bowlful.

2 Remove any woody stems or really tough stalks. Cut the rest into pieces about 5 cm long.

3 Dissolve 30 g of washing soda in cold water and pour it into a large, old saucepan. Do not use an aluminium pan. Fill the pan three-quarters full with water.

4 Add your plant matter. Make sure it is completely covered by the water. Put a lid on the pot and place it on a gas or electric ring. Bring the liquid to the boil, turn down the heat and simmer very slowly. **Ask an adult to help you do this. Watch very carefully to make sure that the pan does not boil dry. Open windows and doors to allow steam to escape.**

5 After three-quarters of an hour, check to see if the plants are ready. Ask an adult to remove a small amount of the plant matter from the pot and squeeze it between their fingers. If it comes apart or feels very soft it is probably ready.

Remember that tough plant matter, such as pineapple tops, will take longer to soften than other matter, such as cabbage leaves and rhubarb stalks.

6 Put a piece of net curtain material over a washing-up bowl in a sink. Pour the liquid and plant matter into the bowl. Pick up the edges of the material and let the liquid run through. Gather up the material to make a small bag and hold it under a running tap. Keep it under the tap until the liquid coming out runs clear.

7 Some plant matter is turned a nasty brown colour by the washing soda. If this happens, put some bleach in a bowl, add cold water, stir well and add the plant matter. The bleach will lighten the colour. After an hour, pour the liquid into a piece of netting and rinse under a tap. **Be careful not to splash the bleach around or on yourself. Ask an adult to help you.**

8 The plant matter now needs to be beaten to make a smooth pulp – just like pulp made from recycled paper. Turn to pages 12-13 for instructions on how to do this. Many plant papers shrink and buckle as they dry. Dry them under weights to keep them as flat as possible.

Always be careful with gas and electric rings. Ask an adult to help you with them.

Making Coloured and Textured Papers

You can add colour and **texture** to your paper in many different ways. Experiment to get the effects you want.

Adding Colour

1 Make a mug of tea using a tea bag. Leave the bag in the water for at least a quarter of an hour and squeeze it with a spoon. Add the tea to your pulp. Ordinary tea will produce various shades of brown. A mug of coffee will produce a slightly darker shade of brown. Camomile tea will make a soft yellow and, surprisingly, fruit teas will make shades of grey.

2 To make bright colours, add some pieces of coloured paper when you make your pulp. The **dye** in it will colour the pulp. Paper napkins give really good colours. You can mix colours together – for example, yellow and red napkins together produce a very good orange colour.

3 To make even brighter colours, mix powder paints with a little water and stir them into your pulp.

4 Try using natural dyes. For example, onion skins boiled in water give a range of soft orange colours. Add the water to your pulp.

▼ *Papers made using different colourings: (clockwise from top left) curry powder; onion skins; tea leaves; instant coffee; (centre) powder paints.*

ADDING TEXTURE

1 When you have made your pulp, add other objects to it. Try flower petals, leaf skeletons, seeds, tiny pieces of ribbon, threads of silk, cotton or wool, pieces of metal foil or scraps of other papers.

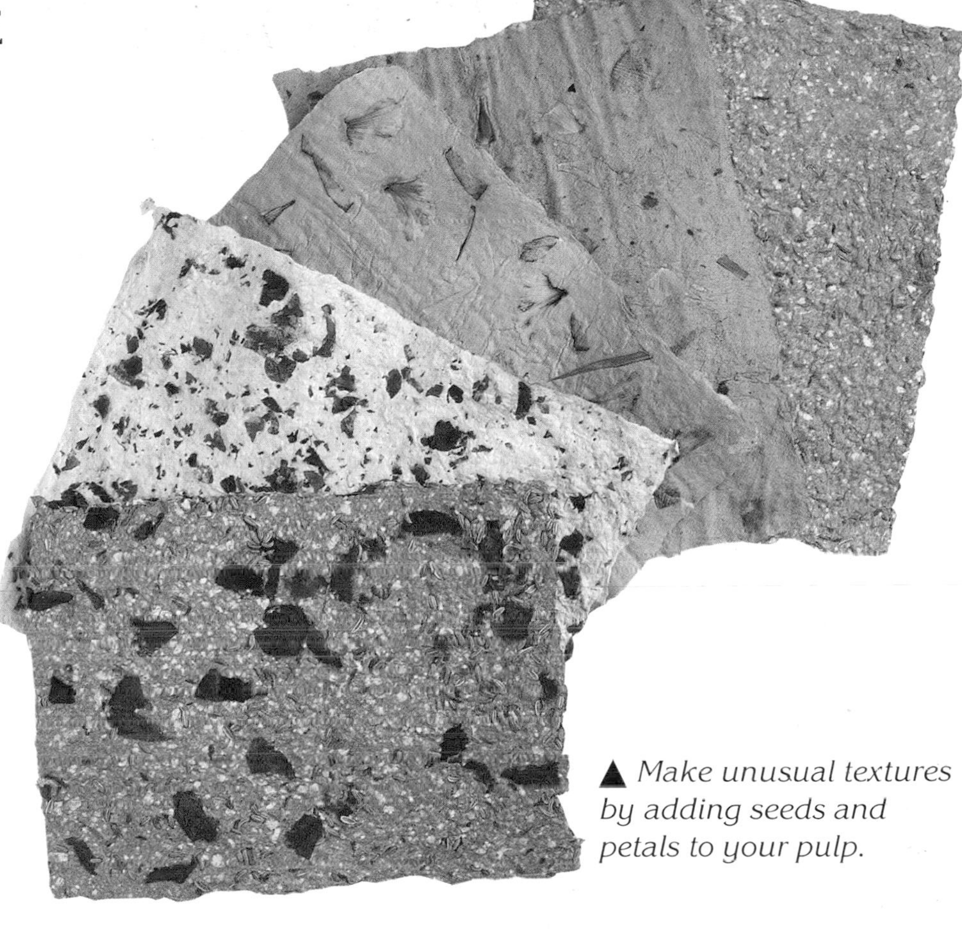

▲ *Make unusual textures by adding seeds and petals to your pulp.*

2 Leave your paper to dry on a textured surface, such as a piece of tweed fabric, or a lace mat.

3 Instead of using a piece of plain net curtain as a mesh on your mould, try a piece of patterned lace.

4 **Stamp** objects into the damp paper to make a raised pattern. This is called **embossing**.

Making a Wall-Hanging

Paper making can be an art in itself. Use what you know about making paper to produce a work of art which has a special message or meaning. For example, if you make pulp from the needles of your Christmas tree, the paper you make will remind you of Christmas. If you recycle an old address book, the finished sheets will have a special meaning for you. Make sure other people understand the meaning of your work by carefully choosing the colour, texture and edges you create. Give your work a title which gives a clue to its message.

1 Spend time thinking about your picture and how you can put across your ideas through paper.

2 Look at the work of other artists – painters, sculptors, photographers and craftspeople. See how they use colour, texture and titles to put across their ideas and feelings.

3 Collect paper samples and make sketches of things you see to give you more ideas. Study the textures of different objects.

4 Decide what materials you need to use, collect them together and make your pulp.

▲ *These hangings are called* Christmas *and* A Summer's Day. *The summer piece is made from garden plants. The Christmas piece is made from wrapping paper and the coloured foil from crackers.*

5 Make your sheet or sheets of paper and leave to dry. Remember that the edges do not necessarily need to be straight.

6 Now hang your work so that it can be displayed. Pin it to a pinboard or mount it on a large sheet of stiff paper.

Making a Paper Bowl

Damp sheets of newly made paper are very flexible. If a sheet is pressed over a solid object, it will take on that shape itself when it dries. This method of making **three-dimensional** objects is called 'casting'.

1 Find a plastic or china bowl of the size you would like to make. You can mould the paper either on the inside or the outside of the bowl.

2 Put a few drops of vegetable oil on a paper tissue and wipe the surface of the bowl. This will make it easier to remove the paper shape later.

3 Make some sheets of paper. If you want to make a large bowl, you will need to use a large picture frame as your mould when you make the sheets.

4 Press several layers of damp paper into the bowl. Use a spray bottle to dampen the sheets of paper with water if necessary.

5 Either fold down the edges of the sheets of paper inside the rim of the bowl, or leave them ragged. Leave to dry out completely.

6 When the paper is dry, slip the moulded paper bowl out of the plastic or china bowl.

FINISHING

There are lots of different ways of finishing the bowl. Here are some ideas.

1 Varnish the paper bowl, inside and out, and leave to dry. Paint on another two or three coats, leaving the varnish to dry between each coat.

2 Brush several coats of vegetable oil on the inside and outside of the bowl.

3 Coat the bowl with wax (see page 17). You can add extra pieces of paper to the surface and rim of the bowl and use the wax to seal them on.

▼ *Make a variety of items using different bowls and plates.*

INVESTIGATING PAPERS

Collect as many different sorts of paper as you can find. You should be able to find examples of **newsprint**, writing paper, airmail paper, white and brown envelopes, greaseproof paper, tracing paper, waxed paper cartons (for soup or milk), cellophane sweet wrappers, Christmas wrapping paper, brown wrapping paper, paper handkerchiefs, tissue paper, a disposable nappy, blotting paper, shiny paper, wallpaper and recycled papers (including some you have made yourself). Try to include some Japanese paper from an art shop and some paper with a **watermark**.

1 **Absorbency** Put a little water in a saucer. Place each paper in turn on the water. Do any mop up the water?

Most of the papers listed above have particular uses. They have different qualities which make them suitable for the job they do. For example, disposable nappies and paper handkerchiefs need to be soft and absorbent. A milk carton must not leak. Writing paper needs a non-absorbent surface to hold the ink. Brown wrapping paper needs to be strong.

Now test your paper samples. Try these experiments and carefully note down your results on a chart.

2 **Translucency** Hold the papers up to the light. Can you see through them?

3 Watermarks Hold the papers up to the light. Can you see a word or pattern?

4 Coating Try writing on the different papers with an ink pen. If the paper is not coated, the ink will sink completely into the surface. If the paper has a lot of coating, the ink will sit on the surface and smudge easily.

5 Rattle Well-made paper has what is known as a good 'rattle'. Hold a piece of paper in one hand and flick it with the fingers of your other hand. Can you hear the different sounds the various papers make?

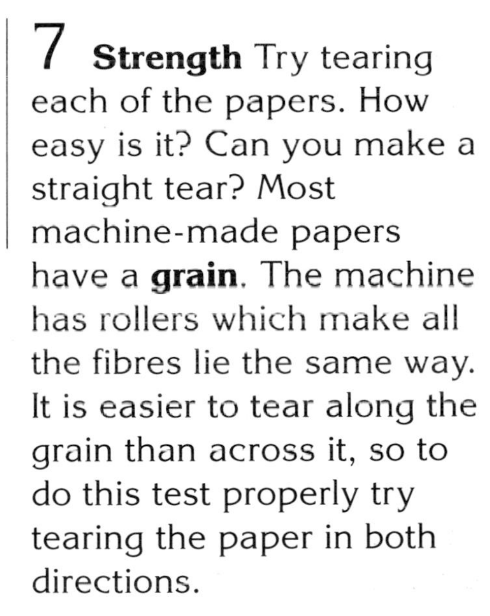

6 Texture Shut your eyes and run your fingers lightly over the different papers. What do the different surfaces feel like?

7 Strength Try tearing each of the papers. How easy is it? Can you make a straight tear? Most machine-made papers have a **grain**. The machine has rollers which make all the fibres lie the same way. It is easier to tear along the grain than across it, so to do this test properly try tearing the paper in both directions.

	blotting paper	wrapping paper	tracing paper	sketchbook paper	recycled paper
hand-made					✓
machine-made	✓	✓	✓	✓	
absorbent	✓			✓	✓
translucent			✓		
watermarked				✓	✓
coated		✓	✓	✓	
uncoated	✓				✓
good rattle		✓		✓	
texture				✓	✓
tears across	✓		✓	✓	✓
tears down		✓	✓		✓

THE GALLERY

Hand-made paper has special qualities. It may have an interesting surface texture, or uneven edges. Other natural and hand-made objects have similar qualities. Look around you. Try to notice edges and rims. Look out for **weathered** objects. Take photographs and collect pictures to stick in a notebook. Use your collection as a source of ideas to try out in your paper making.

These pictures will give you some ideas. The ones on this page show interesting textures. The ones on the opposite page will give you ideas for edges.

◀ *Weathered stone.*

▲ *The skin of citrus fruit.*

▼ *Rusty metal.*

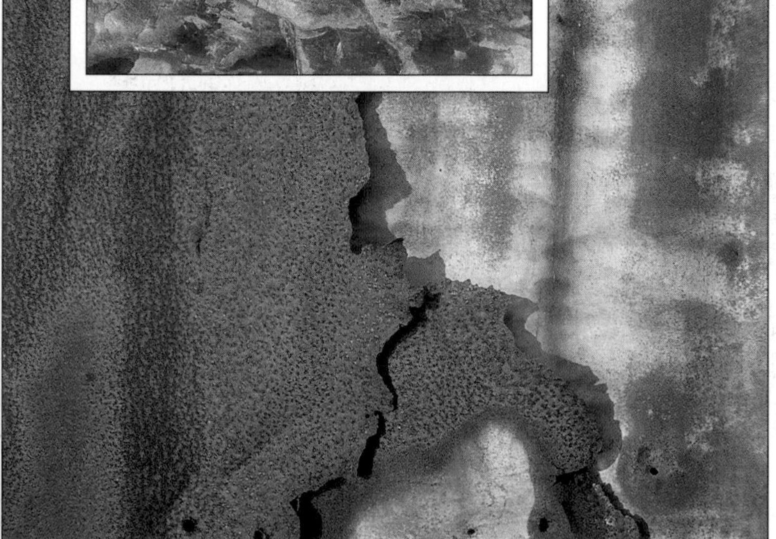

▼ *Sea spray.*

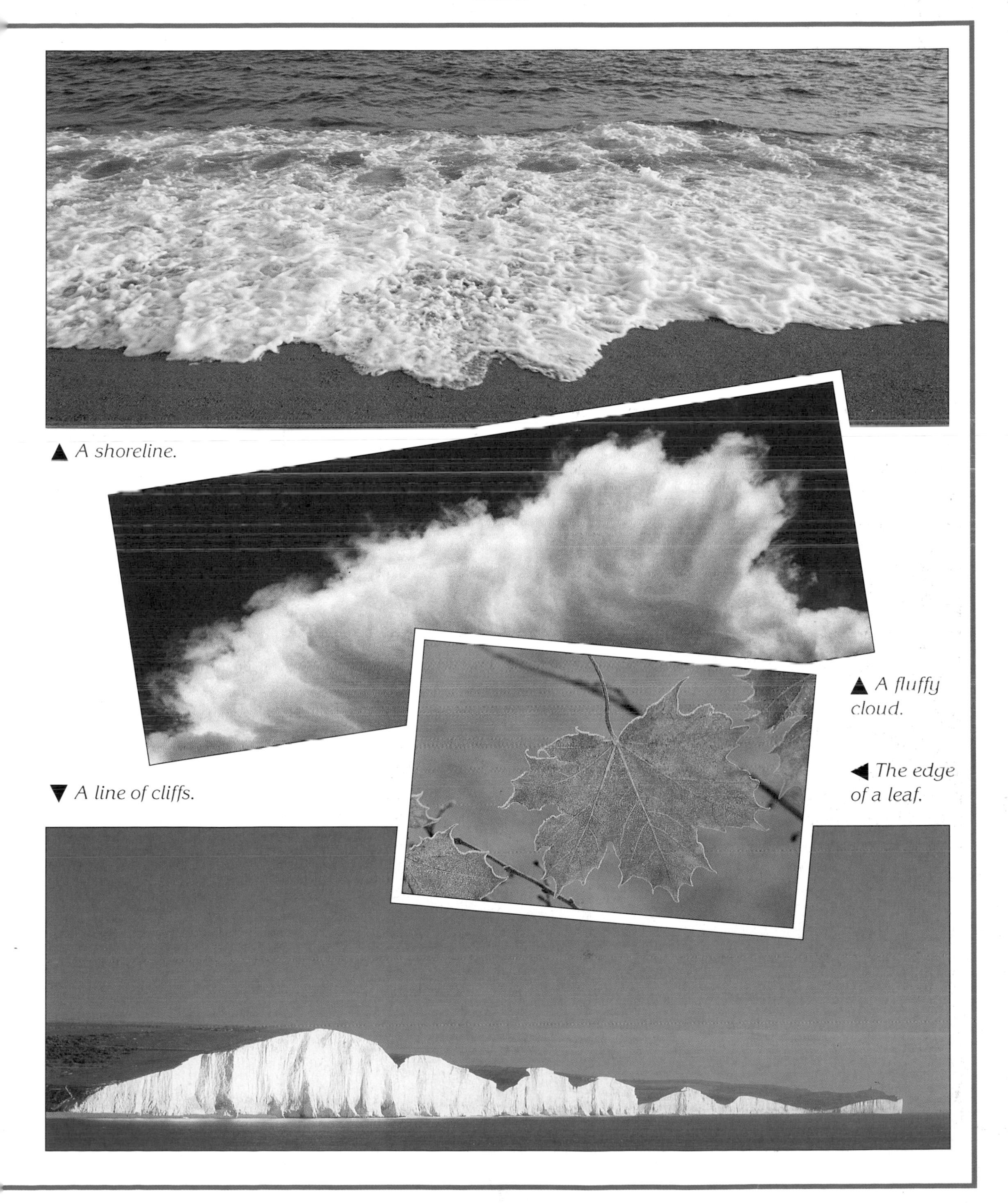

▲ *A shoreline.*

▲ *A fluffy cloud.*

◀ *The edge of a leaf.*

▼ *A line of cliffs.*

GLOSSARY

Atmosphere The air that surrounds the Earth. It is made up of different gases.
Calligraphers People who are very good at writing by hand. To calligraphers, beautiful writing is an art.
Climate The kind of weather that usually occurs in a particular part of the world.
Conifers Trees, such as pines, that have evergreen leaves and produce cones.
Disposable Intended to be thrown away after use.
Dye A substance, usually a powder or a liquid, that is used to colour fabric, paper or other materials.
Embossing Making a raised pattern on a flat surface, usually by pressing something into it.
Environmentalists People who work to protect the environment. The environment is the natural surroundings – on land, in water or in the air – in which plants and animals live.
Fibres Tiny, thin threads which make up the stalks, stems and leaves of plants and the hairs of animals. Plant fibres are used to make paper and other materials.
Finish The surface texture of a material such as cloth or paper.
Flexible Able to bend without breaking.
Grain The general direction of the fibres in a piece of wood or paper.
Hardwood Wood from trees such as oaks and beeches, which has a hard, close-grained texture.
Kimonos Traditional Japanese robes.
Laminated Bonded together into sheets.
Linen Woven fabric made from flax plants.
Loom A frame used in weaving.
Manufacturers People or businesses that make goods, usually on a large scale.
Mould In paper making, the frame on which the sheet of paper is made. It has a fine screen stretched across it which catches the pulp and strains out the water.
Newsprint The special name for the paper that newspapers are printed on.
Non-absorbent Unable to soak up liquids.
Papyrus A tall, reed-like plant.
Pulp A soft, soggy substance.
Recycled Treated and reprocessed to be used again.
Saliva The liquid in an animal's mouth that makes food easier to swallow.
Softwood Wood from trees such as conifers, which has a loose, open-grained texture.
Stamp Press down so as to make a mark.
Tablets Slabs of stone, clay or wax used for writing on before paper was invented.
Taut Extremely tight.
Texture The feel of an object's surface.
Three-dimensional Having three dimensions – height, width and depth.
Translucent Allowing light to show through.
Varnish A clear, sticky liquid which can be painted on an object. When it dries it becomes hard and shiny.
Vat A large container for holding liquids.
Watermark A mark put on paper which can only be seen when it is held up to the light. Watermarks are put on banknotes to make them difficult to copy.
Weathered Worn away by the weather.
Woven Made by weaving – interlacing threads to make a kind of grid.

Further Information

Books to Read

For children:
Burt, Erica *Paper and Card* (Wayland, 1989)
Condon, Judith *Recycling Paper* (Franklin Watts, 1990)
Dixon, Annabelle *Paper* (A & C Black, 1989)
Langley, Andrew *Paper* (Wayland, 1991)
Thomson, Ruth *Making Paper* (Franklin Watts, 1987)

For parents and teachers:
Shannon, Faith *The Art and Craft of Paper* (Mitchell Beazley, 1987)

Useful Addresses

Australia
Pulp and Paper Manufacturers Federation of Australia,
GPO Box 1469N,
Melbourne, Victoria 3001

Canada
Canadian Pulp and Paper Association,
Sun Life Building, 19th Floor,
1155 Metcalfe Street,
Montreal, PQ H3B 4T6

New Zealand
New Zealand Pulp and Paper Industry Association Inc,
PO Box 1650,
Trust Bank Building, 7th Floor,
Hinemoa Street, Rotorua

UK
Pulp and Paper Information Centre,
Papermakers House,
Rivenhall Road, Westlea,
Swindon, SN5 7BE

USA
American Paper Institute,
260 Madison Avenue,
New York 10016

For further information about arts and crafts:

The Crafts Council,
44A Pentonville Road,
London, N1 9BY, UK

Crafts Council of New Zealand,
22 The Terrace,
Wellington,
PO Box 498,
Wellington Island, NZ

Places to Visit

The British Library has a permanent exhibition on paper at:
The British Museum,
Great Russell Street,
London,
WC1B 3DG, UK

The Dard Hunter Museum of Paper,
Massachusetts Institute of Technology,
Cambridge,
Massachusetts, USA

Index

Acknowledgements

The publishers would like to thank the following for allowing their photographs to be reproduced: Bridgeman Art Library 5 (Bradford City Art Gallery and Museums); Cephas 4 bottom right; Eye Ubiquitous 4 top, 9 top, 28 bottom left; Michael Holford 6 left; Tony Stone Worldwide frontispiece (A. Diesendruck), 8 left (A. Cassidy), 28 top left (L. Valder), 28 top right (R. Weller), 28 bottom right (P. Berger), 29 top (J. Garrett), 29 upper middle (D. Wilson), 29 lower middle (R. Siegal), 29 bottom (O. Benn); Topham 7 top, 8 right, 9 bottom right; Wayland Picture Library 4 left (A. Blackburn), 6 right, 7 left, 9 bottom left (A. Blackburn); Zefa 7 bottom right.

All other photographs, including cover, were supplied by Zul Mukhida. Logo artwork throughout the book was supplied by John Yates.

The Diver (one of two) by David Hockney, on page 5, appears by kind permission of the artist: © David Hockney 1978.

The author would like to thank Jan Peek, of Liberty Junior School, Merton, and Jackie Lee and Sarah Mossop, of The Crafts Council, for their help.